Brunna D´Luise T. Lotti Alves

Urban and peri-urban community gardening

Brunna D´Luise T. Lotti Alves

Urban and peri-urban community gardening

Multiple case studies in the city of Campinas

ScienciaScripts

Cover image: www.ingimage.com

This book is a translation from the original published under ISBN 978-613-9-66732-1.

Publisher:
Sciencia Scripts
is a trademark of
Dodo Books Indian Ocean Ltd. and OmniScriptum S.R.L publishing group

120 High Road, East Finchley, London, N2 9ED, United Kingdom
Str. Armeneasca 28/1, office 1, Chisinau MD-2012, Republic of Moldova, Europe
Printed at: see last page
ISBN: 978-620-8-05574-5

Thank you

I would firstly like to thank God for guiding me on this journey, without faith I would have got nowhere.

I would like to thank my supervisor Dr Rosana I. Corraza for her trust in me. Corraza for the trust she placed in me. Her knowledge and dedication were essential in carrying out this work, and especially the affection and attention I received. It was an honour to have you as a teacher.

I would like to thank Mr Orlando Santos for the guided tour of the community garden and for selflessly giving his time to guide me through my project.

Thank you to my parents Altair and Angélica, who have always been by my side, supporting me in every way to encourage me to keep believing in my dreams, even when I disbelieved. Thank you for being there with so much love.

Thank you to my little Beatriz, for being my best friend, for helping me and for being what I am today, because she has been everything and more that a sister should be. Thank you for being my sister, I couldn't have a better family.

I would like to thank my great-aunt Ivone for her patience and wisdom, advising me with her experiences and showing me the way to go. My gratitude goes to Lúcia (Dedé), who saw the value of friendship and commitment with my family.

Thank you to my boyfriend Leonardo, whose care and love helps me at all times, with whom I have learnt the meaning of loving and being loved. Thank you for being my anchor.

I would like to thank Dr Pedro and Dr Teixeira, who showed me that wisdom is the most precious commodity there is, and for being by my side from the very beginning, through all the triumphs and tribulations.

I'd like to thank Professor Marina Garcia, who has encouraged me to do research ever since I was at university, reading and correcting my work. I will always look up to her.

I would like to thank Núria for her friendship over the years, for her literary recommendations and, above all, for being a great friend.

I would like to thank my friends that I made during my degree and that I will carry with me for life: Charles and Rhaysa. They were with me in good times and bad, relieving me of difficulties with tenderness and laughter.

And finally, with great gratitude, I dedicate it to my grandmother Tereza. She introduced me to the magic of study and the importance of dedication. Wherever you are, you will always be in my heart. Thank you for being my reference point.

I dedicate this work to my mother Angélica who supported me so much, to my father Altair for his patience, to my sister Bia for her motivation, to my great-aunt Ivone for her prayers, to my boyfriend Leonardo for his love and affection and also to my grandmother who encouraged me so much to get a degree, may she see me wherever I am.

"A rooster alone cannot weave a morning: he will always need other roosters. One who picks up his cry and throws it to another; another rooster who picks up the cry of a rooster before and throws it to another; and other roosters who cross the sunny threads of their rooster cries with many other roosters, so that the morning, from a tenuous web, is woven between all the roosters."

João Cabral de Melo Neto

Summary

The main aim of this work is to analyse how urban agriculture works, with an emphasis on community gardens, and to see if they are within the logic of the Solidarity Economy. The work also briefly discusses the definition of the concept of urban agriculture based on a theoretical foundation, defining the main advantages and disadvantages of the practice with a case study in Campinas. The specific objective was to understand the practices of self-management of resources at community level, analysing under the principles of solidarity and cooperation, how social and economic relations take place and how they requalify the meanings of work. The methodology consists of two stages. The first stage, a bibliographical review, sought to compile a theoretical framework on Urban and Peri-urban Agriculture and Solidarity Economy. In the second stage, a documentary approach was used to identify community garden initiatives in the city of Campinas, based on secondary sources involving data made available by Campinas City Hall, as well as to characterise the projects according to neighbourhood MHDIs. In the third stage, field research was carried out in Parque Itajai 3, with the aim of gaining a more systematic and in-depth understanding of the principles, practices and results that have been achieved by these initiatives, through technical visits and interviews with the players involved. The results showed the difficulty of research relating Urban Agriculture, Peri-urban Agriculture and Horticulture, with conceptual discoveries and distinctions between concepts being superficial and fieldwork becoming incompatible with the proposal, so that there is a lack of systematised information on vegetable gardens in the municipality of Campinas.

Key words

Urban Agriculture; Urban Community Gardens; Solidarity Economy.

Summary

CHAPTER 1

Introduction to the topic, justification and objectives

1.1 Introduction and justification

At the beginning of the 1970s, in the countries of the South, urban agriculture became the subject of a growing number of interventions by public agencies and non-governmental organisations (NGOs). These interventions can generally be included within social development projects for the poor populations identified as beneficiaries, as part of a dynamic of community economic development (BOULIANNE, 2000).

The cultivation of vegetables in urban and peri-urban areas emerged worldwide from the 1980s onwards, involving initiatives in Latin America, Africa and Asia, partly as a way of tackling the problems experienced in these regions in the face of the economic crisis at the time (CASTELO BRANCO and ALCANTRA, 2011).

For the FAO, the accelerated process of urbanisation has been accompanied by land planning, and as a result of spatial segregation it is estimated that more than 162 million people live in hazardous and degraded areas. The growth of cities has a negative impact.

The relationship between the urban and the rural is capable of generating creative and dynamic forms of occupation and land use, and as a result the city absorbs the countryside (LEFEBVRE, 1978). The urban can be defined as the centrality, relationships and activities developed by "beings conceived, constructed or reconstructed by thought" (LEFEBVRE, 1999, p. 54), while the rural would be the social morphology. For the author, not only is the urban in a relationship of dominance over the rural on a political, economic and cultural level (since it is in the urban environment that the centres of power, the most dynamic elements of capitalist exploitation and information systems are to be found).

Urban sprawl leads to the loss of production areas and food on the outskirts of cities. One consistent fact is that the poor spend more than 50 per cent of their income to buy the food they need (FAO).

The topic of Urban and Peri-urban Agriculture is gaining prominence on the world and national stage, and can be considered a factor in the sustainable development processes of people and society (ARRUDA, 2006).

Current literature refers to Mougeot (2005), who defines urban agriculture based on factors such as: the economic activity; the intra-urban or peri-urban location; the types of areas where it is practised; its scale and production system; the categories and subcategories of products (food and non-food); and the destination of the products, including their commercialisation.

When conceptualising urban and peri-urban agriculture, it is worth highlighting the decisive role of public policies to encourage and implement them, which can favour and promote local development on the outskirts of large cities, with a view to territorial, economic, social and political diversity (MACHADO and MACHADO, 2002).

Some authors suggest that urban and peri-urban agriculture practices can help fight poverty and hunger by generating jobs, income and food. They also argue that these practices can contribute to food security by providing fresh food, generating productive jobs and generating income linked to food production (VILELA and MORAES, 2015).

Urban agriculture has numerous advantages for the city space, such as promoting environmental education, the use of herbal medicines, as well as creating a microclimate and maintaining biodiversity (ROESE, 2016).

A particular look at urban agriculture is offered by the Solidarity Economy, whose objects of analysis prioritise new economic relationships, mediated by associated work and the principles of solidarity and cooperation:

> The counter-hegemonic [sic] character of these actions of resistance to the model of economic development that perpetuates the domination of agrarian elites in rural areas or to the capitalist way of organising the relations of human beings with each other and with nature, undoubtedly contributes (...) it is fundamental to understand, however, that in the daily practices of these groups and organisations, resistance to capitalism and survival within **capitalism are part of the same equation. "** (JOB SCHIMITT and TYGEL, 2009)

It is understood that urban agriculture relates to food production within the urban and peri-urban perimeter, using labour-intensive methods, with human-crop-environment interaction and the facilities of urban infrastructure that provide stability for the workforce, diversified crop production and waste recycling (AQUINO and ASSIS, 2007).

Urban planning for the practice of agriculture should be integrated, not limited to planting crops for food, but including species that contribute to aspects related to biodiversity management (MACHADO E MACHADO, 2002).

The feasibility of utilising idle spaces in cities has been observed, mainly for the benefit of local communities, through initiatives such as gardens, community gardens, orchards and composting, leading to improvements in quality of life that tend to go beyond food production in the outskirts of cities (ROSA, 2011). Depending on the characteristics of the initiative, the benefits involve sociability, a sense of identity and belonging, education and even leisure or entertainment.

Urban agriculture can be seen as an auxiliary agent for social and economic disparities, in the sense that it acts to minimise the numerous problems faced by poor populations in the poorest countries or those with major social inequalities (MONTEIRO, 2007).

Some international agencies that help and support UPAs are UNDP, FAO and UNICEF. (MACHADO and MACHADO, 2011)

At national level, urban agriculture has been part of the food and nutrition security policy since 2003, forming part of the Zero Hunger Programme (ARRAES and CARVALHO, 2015).

In the state of São Paulo, the agency that regulates Urban Agriculture is the IAC, through the Hortalimento Programme (2014), ensured by State Decree No. 50.2333, of May 2005, coordinated by CODEA-GRO, the programme transfers resources to municipalities for the purchase of materials, such as hydroponic greenhouses. (ARRAES and CARVALHO, 2015)

In the municipality of Campinas-SP, Law No. 9549 was introduced in December 1997 to create community gardens in the city, which are still active today. The rationale behind the law is to take advantage of unemployed labour, provide occupational therapy and keep the land clean. (Prefeitura de Campinas, 2016)

The second chapter - Urban and Peri-urban Agriculture: concepts and modalities - attempts to compile the relevant bibliography on Urban and Peri-urban Agriculture, Solidarity Economy and Community Gardens.

The chapter Socio-economic Contexts of Urban and Peri-urban Horticulture Initiatives in Campinas presents the background to the initiatives in the municipality, as well as data on the MHDIs of the neighbourhoods, in order to structure the socio-economic profiles of the locations.

In the fourth chapter - A case study of HUP in Campinas: the Parque do Itajaí vegetable garden - we sought to characterise the IDHMs of the neighbourhood, and carried out fieldwork to obtain detailed studies.

The Final Considerations show the emblems that this work faced, as well as some discussions about urban horticulture in Campinas.

1.2 Objectives

General Objective

The main aim of this work is to identify and analyse Community Garden initiatives in the municipality of Campinas in the light of Solidarity Economy concepts and principles, in order to verify and describe the main advantages perceived by gardeners and the main bottlenecks they have identified within these initiatives.

Specific objectives

The specific objectives of this work will be to analyse community garden initiatives established within the urban and peri-urban perimeters of Campinas in the light of the principles of Solidarity Economy, in order to:

i- Identify concepts and define operational principles of the Solidarity Economy that can be applied to analysing community horticulture initiatives in urban and peri-urban contexts.

ii- Identify the interest and needs of the actors and the search for a new meaning in their relationship with the environment, due to economic, cultural or environmental factors.

iii- Understand the practices of self-management of resources at community level, including the social, technical and organisational aspects.

iv- To analyse, under the principles of solidarity and cooperation, how social and economic relations take place and how they requalify the meanings of work.

v- To identify and describe how the actors involved understand and value the results achieved within the framework of the initiatives studied.

1.3 Methodology

The methodology used in this Course Conclusion Work was conducted in two stages: a literature review and field research, the first of which was carried out in two phases: one exploratory and the other to search for secondary data, while the second, corresponding to fieldwork, collected primary data on a community garden initiative in Campinas, which was analysed, as detailed in figure 1 below.

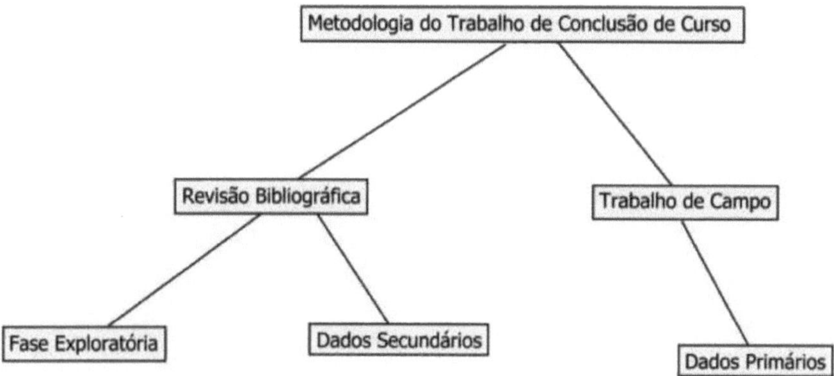

Fig. 1: Flowchart of research activities

Organisation: own elaboration, 2016

First stage

The methodology corresponding to the first stage was a literature review, seeking a theoretical basis in the Urban and Peri-urban Agriculture and Solidarity Economy framework using authors such as Aquino and Assis (2007), Mougeot (2005), Machado (2002), Monteiro (2007), Roese (2016), Rosa (2011), Singer (2008) and others.

The work was based on Juliana Arruda's dissertation on Urban and Peri-urban Agriculture in Campinas-SP: Analysis of the Community Gardens Programme as a subsidy for Public Policies. It was based on secondary data.

Subsequently, an empirical search located community garden initiatives in the city of Campinas through blogs, news articles and the Campinas City Hall website. The identification and characterisation of the initiatives promoted by governmental organisations (Campinas City Hall) and civil society organisations (NGOs, universities), as well as their location, made it possible to draw up the initial information summarised in Table 1, as well as Map xx. Eight cases were compiled in the city of Campinas, distributed throughout the outskirts of the city.

Using the data in the table, it was possible to find the MHDIs (Municipal Human Development Indices) of all the neighbourhoods that contain the initiatives through the Atlas of Human Development in Brazil website, carried out by the UNDP, IPEA and FJP. Because the Campinas region is a Metropolitan Region (MR), it was possible to obtain the values for Education, Health and Income for each locality. There was, however, an obstacle: some of the neighbourhoods are not included in the Atlas, and the author's methodology was to group them according to their location, due to the proximity of the neighbourhoods. The initial steps are described in the Exploratory Phase section below.

Exploratory phase

In the exploratory phase, we sought to define initial concepts such as Urban and Peri-urban Agriculture and differentiate it from Rural Agriculture. Another definition that was sought was Solidarity Economy, in order to understand the mechanism of operation of the initiatives in the city of Campinas.

Within Urban Agriculture, we searched for authors who portray Community Gardens in Brazil and, in a previous search to reinforce the data on gardens, we searched for a website in which a collaborative map seeks to map gardens around the world.

Subsequently, with the address data collected from websites, newspapers, *blogs* and also from Juliana Arruda's dissertation, a map was made using the satellite image of the city of Campinas and the respective addresses, thus geo-referencing the eight community gardens so that their concentration and socio-economic data could be understood. The map was produced using ArcGis 10.3.1 *software*.

Once mapped, the MHDIs of the neighbourhoods were sought in order to understand the socio-economic situation of each initiative.

Secondary Data

The secondary data was obtained from articles and master's theses that were the result of research carried out by two researchers from UNICAMP, at the Faculty of Agricultural Engineering (FEAGRI), Juliana Arruda, in 2005, who was studying for a master's degree at the same university and currently holds a doctorate from UFRRJ, and Professor Dr Nilson Araes, currently a lecturer at FEAGRI.

Second stage

In the second stage, there was field research, the aim of which was to gain a more systematic and in-depth understanding of the principles, practices (social, technological and organisational) and results that have been achieved by these initiatives, through technical visits and interviews with the players involved. A set of preliminary and exploratory questions is presented in the Annex to this project. Its refinement and synthesis should allow for the development of two types of instruments for collecting information in the empirical stage of the research: a script for technical visits to the HUP initiatives in Campinas and a script for interviews. Primary data collection is better defined in the topic below.

Partial activities carried out in the first half of 2016 included interviews, one which resulted in fieldwork at the IAC (Agronomic Institute of Campinas) with Dr Paulo Trani, and the other, participation in a colloquium at Unicamp with Professor Sônia Maria Pessoa Pereira Bergamasco (FEAGRI) on the *subject* of "Family Farming", held in May.

Primary Data

Interviews were used to supplement the documentary information found on websites, newspapers and blogs about community garden projects in the municipality of Campinas, in order to understand the perception of those involved, as the aim of this work.

An interview was held in August at one of the community gardens in Campinas-SP, with the director of the Cio da Terra Association and also with the founder of the Parque Itajaí 3 garden, Mr João Novais.

The aim of this fieldwork was to get closer to the object of study, to get to know the reality of the participants, to accompany the activities and to see the day-to-day life of the farmers in their respective activities.

It is interesting to emphasise that there was no rejection or inhibition on the part of the interviewees in any way, as it encouraged an atmosphere of exchanging experiences, information and ideas.

CHAPTER 2

Urban and peri-urban agriculture: concepts and modalities

2.1 Some concepts used in this work

In the current context of intense urbanisation, urban and peri-urban agriculture is re-emerging around the world as one of the answers to the problems generated by this process, thus contributing to the problem of food shortages (VILELA and MORAES, 2015).

When it comes to concepts of urban and periurban agriculture, they are still a diffuse field, especially when it comes to consolidating public policies. (VILELA and MORAES, 2015; CASTELO BRANCO and ALCANTRA; n.d.)

In literature, since the 19th century, there has been an interaction between agriculture and cities, with the territory interacting with Geographical Science in the work of Elisée Reclus, a French geographer who already correlated natural and human phenomena in all their multidimensionality and complexity. Cities are always in a constant process of (re)creation and end up being understood as a living organism, which, like any organism, must be preserved in order to remain healthy." (ROSA, 2011)

Urban agriculture is carried out in small areas within or around a city, known as peri-urban agriculture, where crops are produced for a variety of purposes, either for their own use and consumption or for small-scale sale in local markets. Both types of agriculture differ from traditional (rural) agriculture in several respects: firstly, the area available for cultivation is very restricted in urban agriculture, in addition to the lack of technical knowledge on the part of the agents and producers directly involved, there is a great diversity of crops and it is often not a requirement of urban agriculture to make a financial profit (ROESE, 2003).

The elements used to define urban agriculture are: the types of economic activities carried out; the categories of products (food and non-food); the type of location (intra-urban and peri-urban); the types of areas where it is practised (public, private); the types of production systems and the destination of the products and the scale of production (MOUGEOUT, 2005).

Below we will use some ideas from Santos (1997), Sposito (2006) and Silva (1998) in order to clarify some of the differences and relationships between the rural and urban dimensions.

The dichotomy between countryside and city is a subject that leads to a number of conflicts, in which the boundaries between the two may not be very clear. This relationship is discussed by Santos (1997), and the definition of countryside and city goes beyond the perception of the landscape or the activities carried out. Rural and urban are categories that refer to an abstract, particular and internal aspect of a given space, in this case, rural and urban fall under the concept of abstract, while countryside and city represent the concrete. In this sense, it's interesting to realise that the categories presented are distinct from each other; both are part of each other's space.

It can be seen that agricultural areas are related to a devouring rationality. It is in this sense that they are more vulnerable than cities, since they are subject to a process of regulation that is commanded by hegemonic market forces, leaving behind precarious forms of local regulation or regulation by public authorities. (SANTOS, 1997)

However, the city is a place that refuses the quick and easy diffusion of new capital, the study of which requires the need to articulate the concept of space, since the city is both a place and a region, since it is a totality and its parts have a combined movement, according to a law of their own. (SANTOS, 1997)

With regard to planning, Santos (1997) states that "In this sense, it can be said that the planning of cities becomes more possible, if not easier, than the planning of agricultural areas. " (SANTOS, 1997, p.67)

Reinforcing this argument, the recognition of an urban/ rural continuum does not mean the disappearance of the city and the countryside as distinct units, but the constitution of areas of transition and contact between these spaces, characterised by the sharing, in the same territory, of land uses, political and economic interests and socio-spatial practices (SPOSITO, 2006).

Arguing about the city/countryside assumption, Sposito (2006, p.126) argues:

> "(which) also goes beyond the material forms themselves, since the link between urban and rural spaces is achieved through means of communication that go beyond, but do not completely overcome, the hierarchical organisation that has governed relations between town and country and between towns in an urban network. Thus, political and economic actors based in the city or the countryside have greater or lesser potential for spatial integration, depending on the means they have to participate in the networks of interaction that are established on the basis of new communication systems."

According to Silva (1998), city-countryside relations used to be less dynamic, but now they are being re-dimensioned. This discussion argues that agricultural regions contain cities and urban regions contain agricultural activities.

An important distinction for the purposes of this work is the recognition of modalities and typologies of urban and peri-urban agriculture. This distinction can be made on the basis of criteria such as: Types of Economic Activity; Type of Area; Location and Type of Production, as we try to point out in the paragraphs below.

In terms of the type of economic activity, urban and peri-urban agriculture are interrelated in time and space, due to the greater geographical proximity and faster flow of resources. The scale of production in this case is not usually large (MOUGEOT, 2005).

With regard to the type of area, what can be noted is that it can be in relation to the producer's residence, inside or outside, as well as the type of land, either by assignment, usufruct, lease, shared, authorised by personal agreement or unauthorised, customary law or commercial transaction, the development of the area can be built up or vacant and the official category of where it is located, residential, industrial, institutional land, etc. (MOUGEOT, 2005).

With regard to the location of activities, it should be noted that there is a conflict in determining where peri-urban areas are located, which is notoriously complex, since their geographical proximity to rural areas means that agricultural transformations are more intense than in the more central and built-up areas of cities. (ROSA, 2011)

However, Roese (2003) disagrees with this perception when he states that urban agriculture is defined by geographical proximity, i.e. it is carried out in small areas within a city, or in its surroundings as in the case of peri-urban agriculture, and one of the differences between this and traditional (rural) agriculture is that the area available for cultivation is very restricted in urban agriculture.

Peri-urban areas are more complex to define than urban areas. It must be close to the city, but the limit can vary from 10 to 90 kilometres, the dependent factor being how it depends on the development of the road infrastructure and transport costs.(MACHADO and MACAHDO, 2002)

Finally, as far as the type of production is concerned, urban and peri-urban agriculture practices focus on the production of crops such as vegetables, spices, aromatic and medicinal plants (TRANI, 2010).

For Machado and Machado (2002) the practice of urban agriculture is related not only to horticulture but also to afforestation, gardens, birds, animals and ornamental plants that are part of urban design.

In the case of the specific object of the study proposed here, namely community gardens, some questions stand out right away, such as: who is responsible for developing them? What is the purpose of the gardens? Who owns the land? And where are they developed? To answer the last question on a global scale, a tool was developed: the collaborative website Falling Fruit, which aims to map community garden initiatives around the world, currently has one million gardens registered, which facilitates the dissemination of knowledge and exchange of experiences.

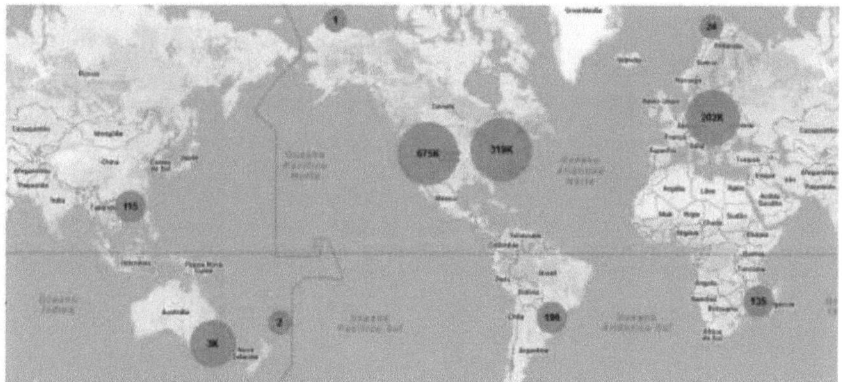

Figura 2 : Distribution of Community Gardens
Source: Falling Fruit, 2016

Regarding the location and type of land, community gardens can be located next to popular restaurants and in vacant spaces (for example, under power lines or next to roads and waterways), or in institutional spaces such as schools, hospitals and businesses. (FAO)

Installation usually takes place on urban land, squares or public areas provided by the town hall, for example, in areas of up to 2,000 square metres (TRANI, 2010).

The activities carried out in community gardens are done by different actors, whether they are neighbours or families from the same community. The tasks are divided up between the participants, who generally work together. The vegetables, medicinal and aromatic plants produced in this type of garden are shared between the participants (TRANI, 2010).

The management of community gardens includes the active participation of the community, which is responsible for the administration and management of the gardens, and the division of labour, which may or may not have the technical support and supervision of the public authorities, as in the case of town halls (ARRUDA, 2005).

The purpose of community gardens would be to benefit local development, in which local production of food or even other types of plants such as ornamental and medicinal plants is valued, and ends up strengthening the solidarity economy and popular culture. Another purpose is food security, which favours total control of all the production phases, a reduction in poverty for own or community consumption, with possible sales of surpluses. (ARRUDA, 2005)

It also acts as an occupational activity, keeping people busy and avoiding idleness, contributing to social and environmental education and reducing the marginalisation of these people in society (ROESE, 2003).

From the perspective of many of the readings carried out for this work, it is not uncommon for urban and peri-urban horticulture initiatives to be associated with Solidarity Economy practices. From a social point of view, community gardens, which are based on a self-management model involving the administration of production by the workers themselves, organised collectively and democratically, contribute to improving social relations, strengthening solidarity, mutual trust and shared decision-making, thus enabling community building (SINGER, 2008; MARTIN, 2015).

The term "Solidarity Economy" has made it possible to bring together various experiences in a single field of ideas, involving various organisations, institutions and people around common objectives. (MOTTA, 2004) Job Schimitt and Tygel (p.108; 2009) set out some visions of Solidarity Economy that are worth highlighting:

> (i) valuing the work, knowledge and creativity of human beings, affirming their supremacy over capital; (ii) identifying associated labour and associative ownership of the means of production as fundamental elements in the construction of renewed forms of economic organisation, based on democracy, solidarity and cooperation; (iii) the democratic management of enterprises by the workers themselves (self-management); (iv) the construction of solidarity collaboration networks as a way of integrating the different enterprises.

The struggle to build a sustainable society is reflected in the construction of another economy, based on new ethics and mediating new labour relations and management of the means of production. So the solidarity economy ends up redefining farmers' relationships with the environment, for example, so that their relationships with the market change. These new relationships, both social and economic, are present in the gardens, which reclassify the meanings of work, productivity, consumption and exchange, strengthening practices of reciprocity and contradicting the idea that human beings seek only profit. (JOB SCHIMITT and TYGEL, 2009)

There is therefore a need to build a new economy, consolidating new ethics and new working relationships, involving practices that differ from purely commercial logic. Thus, community agriculture in urban and peri-urban contexts, made possible by the adoption of vegetable gardens, points to a social and technological change guided by other principles. It is a strategic issue for the environmental and economic dimension, with other social segments on issues such as quality of life, climate change and environmental risks. (JOB SCHIMITT and TYGEL, 2009)

In short, the solidarity economy is one of the pillars of the development model that perpetuates elite domination, and the victims of the crisis are seeking insertion into social production through collective work, such as self-management and participatory administration (SCHIMITT and TYGEL, 2009; SINGER, 2004)).

CHAPTER 3

Socio-economic Contexts of Urban and Peri-urban Horticulture Initiatives in Campinas

The chapter aims to characterise the initiatives according to the socio-economic data of the neighbourhoods in question, in order to justify the presence of the gardens in these locations. As can be seen on the map below, community gardens are scattered throughout the municipality of Campinas, which made it difficult to carry out more fieldwork, but it was possible to find out about one in Parque Itajaí 3, as will be shown in the next chapter.

In the first stage of the research that gave rise to this paper, i.e. bibliographical research in periodical databases and news clippings, it was possible to gather information on eight urban and peri-urban horticulture initiatives in the municipality of Campinas.

The second step was to systematise the programmes by their location and Municipal Human Development Index (MHDI). Below is a table identifying the indices for the years 2000 and 2010 respectively, followed by a detailed explanation of each initiative.

Initiatives	Neighbourhood	MHDIs 2000	MHDIs 2010
1	Campo dos Amarais	0.653	0.766
2	Forest Park 3	0.508	0.636[1]
3	Jardim Santa Rosa	0.608	0.718
4	Eucalyptus Park	0.778	0.855
5	Itajaí Park 3	0.508	0.636
6	Florence Garden	0.608	0.718[2]
7	Vila Brandina	0.564	0.706
8	St Mark's	0.608	0.731

Table xx: Summary of the MHDIs of the neighbourhoods containing the initiatives

Information on these initiatives is detailed in Table xx below, which shows: Name of Initiative, Address, Neighbourhood/Region, Actors Involved, Observations and Miscellaneous, and Source of Information on the Internet.

Table 1 - Urban and peri-urban gardens in Campinas: summary of information on the initiatives surveyed

n	Name of the initiative	Address	Neighbourhood/Region	Actors involved	General and Diverse Observations	Information sources / websites
1	Vila Esperança Healthy Community Association	Campinas Avenida Uriassu de Assis Batista - Amarais Region (Amarais Region - Healthy Community Association)	Campo dos Amarais/ North	Association Vila Esperança Healthy Community, City Hall and Unicamp (guiding cultivation)	The gardeners receive support and technical guidance from the City Council. This is a vegetable garden in a degraded area.	http://www.campinas.sp.gov.br/news-integra.php?id=12686

[1] Due to the proximity of the Parque Floresta 3 and Parque Itajaí neighbourhoods, the MHDI of the latter neighbourhood was adopted, as this information is not available for the former.

[2] The same applies to the neighbourhoods of Jardim Florence/Liliza and Jardim Santa Rosa, because of their proximity, the MHDI was adopted for the second neighbourhood.

2	Seo Benicio's vegetable garden in Forest Park 3	Rua Juvenal Fernandes, Parque Floresta, Campinas	Park Forest/North-West	No information	The group that takes care of the area by planting various species of food plants is now made up of seven residents of the neighbourhood.	http://aproagriup.blogspot.com. br/
3	Santa Rosa Production Centre	Rua Manuel Isidoro Reis, Jardim Santa Rosa- Campinas	Jardim Santa Rosa/ Northwest	Cio da Terra Association	Cultivation in open and protected areas.	http://aproagriup.blogspot.com. br/
4	Eucalyptus Park community garden	London Garden	London Garden/ North West	No identification	The garden is not for commercial purposes and the produce is for their own consumption and the surplus is donated to friends, relatives or families. According to an interview, this garden is the first initiative in the city of Campinas, created in 1983.	http://aproagriup.blogspot.com. br/
5	Parque Itajaí Community Garden 3	Campinas (Rua Doutor Pedro Miguel - Parque Itajaí 6 (one block from Terminal Itajaí))	Itajaí/ Northwest	City Hall and Ceasa	The garden was created with the aim of serving the community. Today, production is not limited to self-consumption, but also to commercialisation and donation to needy communities.	http://aproagriup.blogspot.com. br/
6	Community garden at Jesus Christ the Liberator Parish	Campinas (Rua: Carlos Roberto Pereira, 842. Jardim Florence II, Liliza Campinas, SP). The vegetable garden is located in Praça 03. .	Campo Grande/ Florence/ North West	Casa Maria de Nazaré, Campinas City Hall (PHC)	It emerged as an alternative to environmental degradation in an area of the neighbourhood (ARRUDA, 2006)	
7	Vegetable garden of the NGO Planting Peace on Earth	Campinas Rua Francisco Mesquita, n°1	Vila Brandina/ East	NGO Planting Peace on Earth		
8	Campinas Community Gardens Programme at the Sewage Treatment Plant (ETE) - SÃO MARCOS	R. André Grabois, s/n - Lot. Vila Esperanca, Campinas - SP	São Marcos/ North	Ceasa, City Hall	Hydroponic plants.	http://www.ceasacampinas.com.br/new/DicasVer.asp?id=399& page=tip of the week

Source: own elaboration based on survey data.

It is possible to locate the initiatives under study using the map below, where they are located within the area of the Municipality of Campinas.

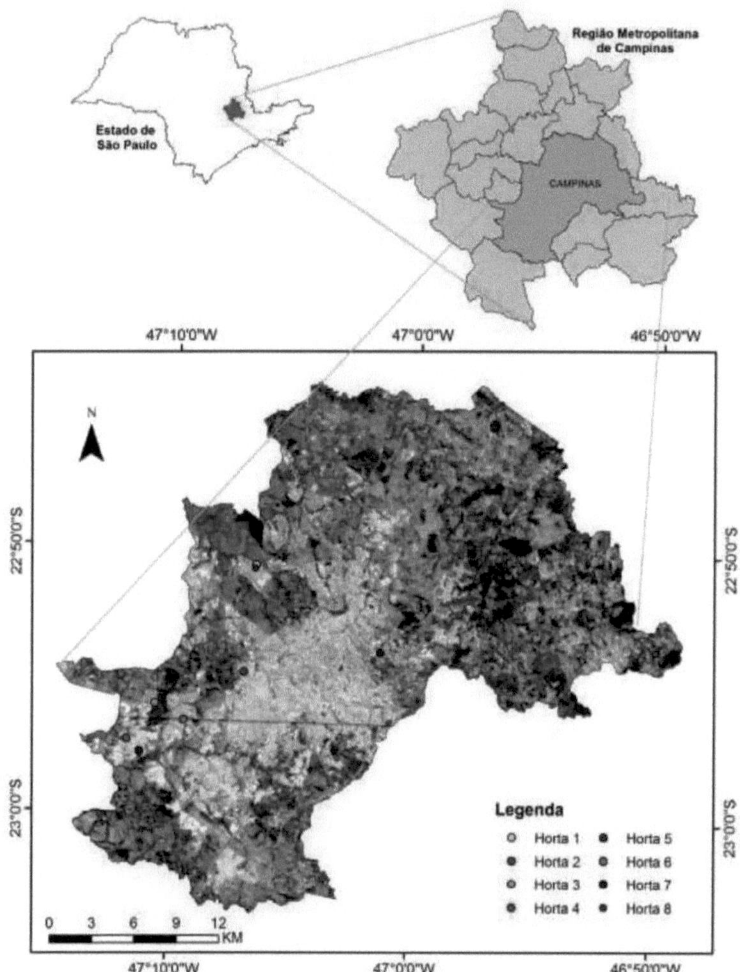

Figura 3 Map of initiatives in Campinas Source: Own elaboration

1.1. NITIATIVE 1 - Vila Esperança Healthy Community Association CAMPO DOS AMARAIS

The Campo dos Amarais region, where the first initiative analysed is located, is in the municipality of Campinas. The MHDI for this neighbourhood was 0.766 in 2010. This value places it in the High Human Development range (MHDI between 0.700 and 0.799). The dimension that contributes most to the UDH's MHDI value is Longevity, with an index of 0.850, followed by Income, with an index of 0.748, and Education, with an index of 0.707. (UNDP, IPEA; 2016)

MHDI and Components	2000	2010
HDI Education	0,518	0,707
% aged 18 or over with complete primary education	42,63	61,35
% of children aged 5 to 6 attending school	65,89	95,84
% of children aged 11 to 13 attending the final years of primary school	81,37	89,94
% of young people aged 15 to 17 with complete primary education	47,16	60,06
% aged 18 to 20 with completed secondary education	34,46	57,21
HDI Longevity	0,774	0,85
Life expectancy at birth (in years)	71,46	76,01
MHDI Income	0,696	0,748
Per capita income (in R$)	608	839,14

Source: Adapted from Source: UNDP, Ipea and FJP

When analysing longevity, the data found in Campos dos Amarais is 13.2 per thousand live births in 2010. The infant mortality rates for the municipality and the metropolitan region are 11.8 and 12.6 per thousand live births, respectively, for the same year. (UNDP, IPEA; 2016)

Life expectancy at birth is the indicator used to make up the Longevity dimension of the Municipal Human Development Index (MHDI). In 2010, while life expectancy at birth was 76.0 years in the UDH, it was 76.6 years in the municipality and 76.5 years in the Campinas metropolitan region (UNDP, IPEA; 2016).

With regard to education in the region studied, the proportion of children aged 5 to 6 in school was 95.84% in 2010. In the same year, the proportion of children aged 11 to 13 attending the final years of primary school was 89.94 per cent; the proportion of young people in school was 95.84 per cent.

of 15 to 17 year olds with completed primary education is 60.06 per cent; and the proportion of 18 to 20 year olds with completed secondary education is 57.21 per cent (UNDP, IPEA; 2016).

The Education MHDI also includes an indicator of the adult population's schooling, the percentage of the population aged 18 or over with completed primary education (UNDP, IPEA; 2016).

In 2010, considering the population aged 25 and over in the UDH, 4.67 per cent were illiterate, 56.29 per cent had completed primary school, 38.67 per cent had completed secondary school and 6.23 per cent had completed higher education. (UNDP, IPEA; 2016)

The average per capita income in the Campo dos Amarais neighbourhood was R$839.14 in 2010, while in the municipality of Campinas it was 1,390.83. In the same year, the proportion of poor people, i.e. with a per capita household income of less than R$140.00 (at August 2010 prices) is 1.47% in the UDH, and 3.16% in the municipality. (PNUD, IPEA; 2016)

Income and Poverty- Campo dos Amarais- Campinas- SP	2000	2010
Per capita income (in R$)	608,04	839,14
% extremely poor	1,41	-
% poor	6,08	1,47

Source: Adapted from UNDP, Ipea and FJP (2016)

In 2010, the activity rate of the population aged 18 and over, i.e. economically active people, was

74.07% in the region. At the same time, the unemployment rate in the UDH (i.e. the percentage of the economically active population aged 18 and over who are unemployed) is 6.38% (UNDP, IPEA; 2016).

Composition of the population aged 18 and over - 2010

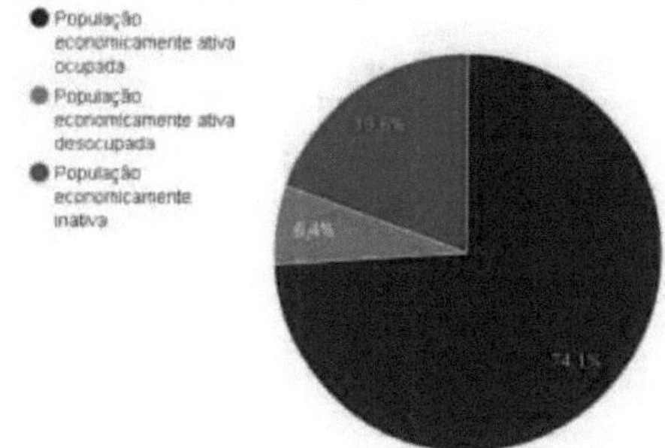

Source: UNDP, IPEA, FJP, 2016

Figure 4: Composition of the population aged 18 and over - Campos do Amarais

Occupation of the population aged 18 and over	2000	2010
Activity rate	64,8	74,07
Unemployment rate	17,28	6,38
Degree of formalisation of the employed - 18 years and over	71,49	76,42
Educational level of the employed		
% of employed with complete primary education	54,1	69,72
% of employed with completed secondary education	28,5	50,96
Average yield		
% of employed people with an income of up to 1s.m.	14,64	9,39
% of employed people with an income of up to 2 monthly instalments	61,09	56,61
Percentage of employed people earning up to 5 minimum wages	93,97	91,31

Source: Adapted from UNDP, Ipea and FJP ; 2016

3.2. INITIATIVE 2- Seo Benicio's vegetable garden in Parque Floresta 3

Forest Park 3

As the Parque Floresta 3 and Itajaí neighbourhoods border each other, there is no data on the MHDI of Parque Floresta, but a field visit revealed that these neighbourhoods have similar socio-economic conditions.

3.3. INITIATIVE 3 - Santa Rosa Production Centre

SANTA ROSA GARDEN

The MHDI of the Santa Rosa neighbourhood, located in the municipality of Campinas, was 0.718 in 2010. This value places the UDH in the High Human Development range (MHDI between 0.700 and 0.799). The dimension that contributes most to the UDH's MHDI value is Longevity, with an index of 0.800, followed by Income, with an index of 0.707, and Education, with an index of 0.655. (UNDP, IPEA; 2016)

MHDI and Components	**2000**	**2010**
HDI Education	0,472	0,655
% aged 18 or over with complete primary education	37,54	56,45
% of 5 to 6 year olds attending school	67,63	89,9
% of 11 to 13 year olds attending the final years of primary school	76,87	85,41
% of 15-17 year olds with complete primary education	42,49	64,68

% aged 18 to 20 with completed secondary education	24,66	42,12
HDI Longevity	0,74	0,8
Life expectancy at birth (in years)	69,41	73,02
MHDI Income	0,645	0,707
Per capita income (in R$)	443,86	649,34

Source: Adapted from UNDP, Ipea and FJP ; 2016

The Longevity MHDI indicates that the municipality's infant mortality rate is 11.8 per 1,000 live births for the same year (UNDP, IPEA; 2016).

In terms of education, the proportion of children aged 5 to 6 in school in the neighbourhood was 89.90% in 2010. In the same year, the proportion of children aged 11 to 13 attending the final years of primary school was 85.41%; the proportion of young people aged 15 to 17 with completed primary school was 64.68%; and the proportion of young people aged 18 to 20 with completed secondary school was 42.12%. (UNDP, IPEA; 2016)

The MHDI for Education in the Adult Population in 2010 was that 7.99% were illiterate, 49.88% had completed primary school, 29.33% had completed secondary school and 3.95% had completed higher education (UNDP, IPEA; 2016).

The average per capita income in Jardim Santa Rosa was R$649.4 in 2010, while in the municipality of Campinas it was R$1,390.83. In the same year, the proportion of poor people, i.e. those with a per capita household income of less than R$140.00 (at August 2010 prices) was 3.84% in the region. (UNDP, IPEA; 2016)

Income and Poverty- Jardim Santa Rosa- Campinas- SP	**2000**	**2010**
Per capita income (in R$)	443,86	649,34
% extremely poor	2,79	1,15
% poor	11,37	3,84

Source: Adapted from UNDP, Ipea and FJP ; 2016

In 2010, the activity rate of the economically active population was 71.20 per cent in the region. At the same time, the unemployment rate in the UDH (i.e. the percentage of the economically active population aged 18 and over who are unemployed) is 7.24 per cent (UNDP, IPEA; 2016).

Composition of the population aged 18 and over - 2010

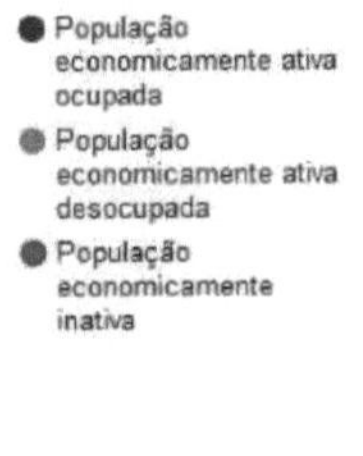

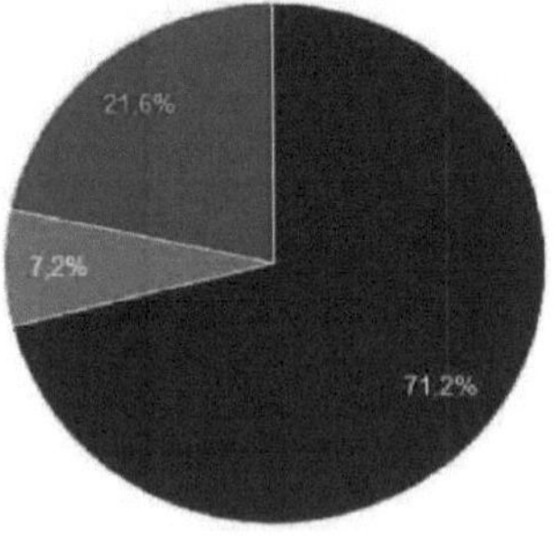

Figure 5: Composition of the population aged 18 and over - Santa Rosa Source: UNDP, IPEA, FJP, 2016

Occupation of the population aged 18 and over	**2000**	**2010**
Activity rate	71,14	71,2
Unemployment rate	15,65	7,24
Degree of formalisation of the employed - 18 years and over	66,63	77,31
Educational level of the employed		
% of employed with complete primary education	44,73	60,72
% of employed with completed secondary education	20,76	36,98
Average yield		
% of employed people with an income of up to 1s.m.	18,72	11,15
% of employed people with an income of up to 2 monthly instalments	70,9	67,02
Percentage of employed people earning up to 5 minimum wages	96,36	95,79

Source: Adapted from UNDP, Ipea and FJP ; 2016

3.4. INITIATIVE 4 - Eucalyptus Park community garden

London Garden

The (MHDI) of initiative 4, located in Jardim Londres in question, located in the municipality of Campinas in the RM of Campinas, is 0.855 in 2010. This value places the UDH in the Very High Human Development range (MHDI between 0.800 and 1). The dimension that contributes most to the UDH's MHDI value is Longevity, with an index of 0.901, followed by Income, with an index of 0.841, and Education, with an index of 0.824.

MHDI and components	**2000**	**2010**
HDI Education	0,71	0,824
% aged 18 or over with complete primary education	62,89	79,67
% of 5 to 6 year olds attending school	85,64	97,94
% of 11 to 13 year olds attending the final years of primary school	74,96	84,94
% of 15-17 year olds with complete primary education	81,12	87,39
% aged 18 to 20 with completed secondary education	60,23	64,85
HDI Longevity	0,84	0,901
Life expectancy at birth (in years)	75,41	79,07
MHDI Income	0,789	0,841
Per capita income (in R$)	1.088,28	1.498,68

Source: Adapted from UNDP, Ipea and FJP; 2016

The longevity MHDI indicates that infant mortality in the neighbourhood was 9.2 per thousand live births in 2010. The infant mortality rates for the municipality and the metropolitan region are 11.8 and 12.6 per thousand live births, respectively, for the same year. (UNDP, IPEA; 2016)

Life expectancy at birth is the indicator used to make up the Longevity dimension of the Municipal Human Development Index (MHDI). In 2010, while life expectancy at birth was 79.1 years in the UDH, it was 76.6 years in the municipality and 76.5 years in the metropolitan region (UNDP, IPEA; 2016).

In relation to education, the proportion of children and young people attending or having completed certain cycles in Jardim Londres, refers to the proportion of children aged 5 to 6 in school being 97.94% in 2010. In the same year, the proportion of children aged 11 to 13 attending the final years of primary education was 84.94%; the proportion of young people aged 15 to 17 with completed primary education was 87.39%; and the proportion of young people aged 18 to 20 with completed secondary education was 64.85%. (UNDP,

IPEA; 2016)

The MHDI for adult schooling in 2010 was 79.67% in the neighbourhood and 67.71% and 64.16% in the municipality and metropolitan region in which it is located, respectively (UNDP, IPEA; 2016).

In 2010, considering the population aged 25 and over in the region, 0.86% were illiterate, 78.62% had completed primary school, 66.57% had completed secondary school and 24.45% had completed higher education. (UNDP, IPEA; 2016)The average per capita income in the UDH was 1,498.68 in 2010, while in the municipality of Campinas it was 1,390.83 and in the Metropolitan Region of Campinas it was 1,148.94. In the same year, the proportion of poor people, i.e. those with a per capita household income of less than R$140.00 (at August 2010 prices) was 0.16% in the neighbourhood. (UNDP, IPEA; 2016)

Income, Poverty and Inequality - Jardim Londres	2000	2010
Per capita income (in R$)	1.088,28	1.498,68
% extremely poor	0,28	0,09
% poor	0,83	0,16

Source: Adapted from UNDP, Ipea and FJP; 2016

In 2010, the economically active population rate was 69.83 per cent in the region. At the same time, the unemployment rate in the UDH (i.e. the percentage of the economically active population aged 18 and over who are unemployed) is 4.77 per cent (UNDP, IPEA; 2016).

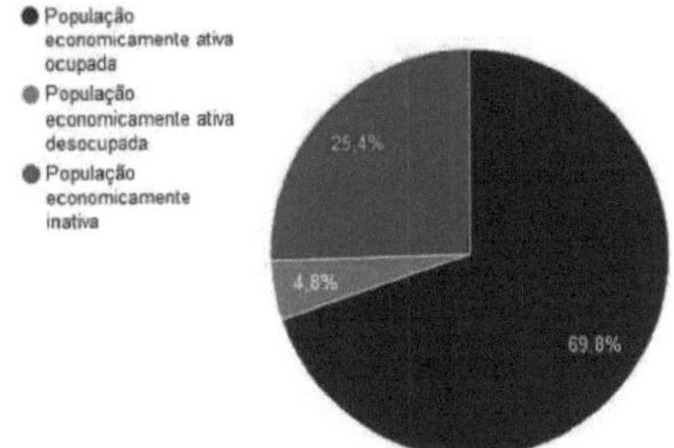

Source: UNDP, Ipea and FJP ; 2016

Figura 6: Composition of the population aged 18 and over - Jardim Londres

Occupation of the population aged 18 and over - Jardim Londres	2000	2010
Activity rate	67,65	69,83
Unemployment rate	14	4,77
Degree of formalisation of the employed - 18 years and over	74,99	81,34
Educational level of the employed		
% of employed with complete primary education	77,03	86,78
% of employed with completed secondary education	57,26	75,15
Average yield		
% of employed people with an income of up to 1s.m.	8,51	4,57
% of employed people with an income of up to 2 monthly instalments	37,13	31,8
Percentage of employed people earning up to 5 minimum wages	73,78	76,41

Source: Adapted from UNDP, Ipea and FJP ; 2016

1.1. INITIATIVE 6 - Community garden at Jesus Christ the Liberator Parish

Florence Garden

The Jardim Florence and Jardim Santa Rosa neighbourhoods also border each other, and the Atlas of Human Development does not include data from the former, but a field visit revealed that both neighbourhoods have similar socio-economic conditions.

3.6. INITIATIVE 7 - Vegetable garden run by the NGO Planting Peace on Earth.

VILA BRANDINA - Northwest

Rua Francisco Mesquita, n°1. Vila Brandina neighbourhood. Eastern region of the municipality.

The Municipal Human Development Index (MHDI) for Vila Brandina in Campinas was 0.706 in 2010. This value places the UDH in the High Human Development range (MHDI between 0.700 and 0.799). The dimension that contributes most to the UDH's MHDI value is Longevity, with an index of 0.787, followed by Income, with an index of 0.694, and Education, with an index of 0.643. (UNDP, IPEA; 2016)

MHDI and Components	**2000**	**2010**
HDI Education	0,391	0,643
% aged 18 or over with complete primary education	28,59	52,74
% of 5 to 6 year olds attending school	63,03	91,11
% of 11 to 13 year olds attending the final years of primary school	61	90,64
% of 15-17 year olds with complete primary education	37,68	64,51
% aged 18 to 20 with completed secondary education	21,32	38,23
HDI Longevity	0,727	0,787
Life expectancy at birth (in years)	68,61	72,22
MHDI Income	0,632	0,694
Per capita income (in R$)	407,78	599,37

Source: Adapted from UNDP, Ipea and FJP ; 2016

The HDI for longevity indicates that infant mortality in the neighbourhood was 19.4 per thousand live births in 2010. The infant mortality rates for the municipality and the metropolitan region are 11.8 and 12.6 per thousand live births, respectively, for the same year. (UNDP, IPEA; 2016)

Life expectancy at birth is the indicator used to make up the Longevity dimension of the Municipal Human Development Index (MHDI). In 2010, while life expectancy at birth was 72.2 years in the UDH, it was 76.6 years in the municipality and 76.5 years in the metropolitan region (UNDP, IPEA; 2016).

Proportions of children and young people attending or having completed certain cycles indicate the school attendance situation among the school-age population and are one of the components of the Education MHDI. In the UDH, the proportion of children aged 5 to 6 in school was 91.11 per cent in 2010. In the same year, the proportion of children aged 11 to 13 attending the final years of primary school was 90.64 per cent; the proportion of young people aged 15 to 17 with completed primary school was 64.51 per cent; and the proportion of young people aged 18 to 20 with completed secondary school was 38.23 per cent. (UNDP, IPEA; 2016)

The Education MHDI also includes an indicator of the adult population's level of education. In 2010, considering the population aged 25 and over in the UDH, 16.76% were illiterate, 46.69% had completed primary school, 25.11% had completed secondary school and 2.61% had completed higher education (UNDP, IPEA; 2016).

The average per capita income in Vila Brandia was R$599.37 in 2010, while in the municipality of Campinas it was R$1,390.83. In the same year, the proportion of poor people, i.e. with a per capita household income of less than R$140.00 (at August 2010 prices) is 5.49% in the UDH, and 3.16% in the municipality. (PNUD, IPEA; 2016)

Income and Poverty- Vila Brandina- Campinas- SP	**2000**	**2010**
Per capita income (in R$)	407,78	599,37
% extremely poor	4,31	1,43
% poor	12,92	5,49

Source: Adapted from Source: UNDP, Ipea and FJP

In 2010, the activity rate of the population aged 18 and over (i.e. the percentage of this population that is economically active) was 73.22 per cent in the Vila Brandina region. At the same time, the unemployment rate in the UDH (i.e. the percentage of the economically active population aged 18 and over who are unemployed) is 10.51 per cent (UNDP, IPEA; 2016).

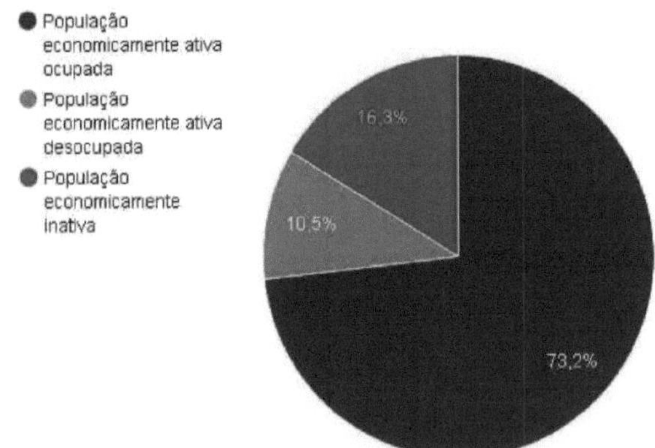

Source: UNDP, IPEA, FJP, 2016
Figure 7 : Composition of the population aged 18 and over - Vila Brandina

Occupation of the population aged 18 and over	**2000**	**2010**
Activity rate	73,07	73,22
Unemployment rate	18,48	10,51

Degree of formalisation of the employed - 18 years and over	68,59	77,66
Educational level of the employed		
% of employed with complete primary education	33,46	60,05
% of employed with completed secondary education	15,54	34,95
Average yield		
% of employed people with an income of up to 1s.m.	20,36	7,16
% of employed people with an income of up to 2s. m.	72,91	72,28
Percentage of employed people earning up to 5 minimum wages	97,54	97,93

Source: Adapted from UNDP, Ipea and FJP. 2016.

3.7. INITIATIVE 8 - Campinas Community Gardens Programme at the Sewage Treatment Plant (ETE) - SÃO MARCOS

SAINT MARCOS

The MHDI of the São Marco neighbourhood, located in the municipality of Campinas, was 0.731 in 2010. This value places the UDH in the High Human Development range (MHDI between 0.700 and 0.799). The dimension that contributes most to the UDH's MHDI value is Longevity, with an index of 0.812, followed by Income, with an index of 0.701, and Education, with an index of 0.685. (UNDP, IPEA; 2016)

MHDI and its components	2000	2010
HDI Education	0,472	0,685
% aged 18 or over with complete primary education	37,54	60,2
% of children aged 5 to 6 attending school	67,63	88,36
% of children aged 11 to 13 attending the final years of primary school	76,87	91,49
% of young people aged 15 to 17 with complete primary education	42,49	66,48
% of young people aged 18 to 20 with completed secondary education	24,66	46,14
HDI Longevity	0,74	0,812
Life expectancy at birth (in years)	69,41	73,71
MHDI Income	0,645	0,701
Per capita income (in R$)	443,9	627,28

Source: Adapted from UNDP, Ipea and FJP. 2016.

The Longevity MHDI shows that infant mortality in the region was 16.8 per thousand live births in 2010. The municipality's infant mortality rates were 11.8 per thousand live births for the same year (UNDP, IPEA; 2016).

With regard to education, in the São Marcos neighbourhood, the proportion of children aged 5 to 6 in school was 88.36% in 2010. In the same year, the proportion of children aged 11 to 13 attending the final years of primary school was 91.49%; the proportion of young people aged 15 to 17 with completed primary school was 66.48%; and the proportion of young people aged 18 to 20 with completed secondary school was 46.14%. (UNDP, IPEA; 2016)

The Education MHDI for the adult population in 2010 was 5.30% illiterate, 55.02% had completed primary school, 31.06% had completed secondary school and 3.87% had completed higher education. (UNDP, IPEA; 2016)However, the average per capita income in the São Marcos neighbourhood was R$ 627.28 in 2010, while in the municipality of Campinas it was R$ 1,390.83. In the same year, the proportion of poor people, i.e. those with a per capita household income of less than R$140.00 (at August 2010 prices) was 7.04% in the neighbourhood. (UNDP, IPEA; 2016)

Income and Poverty- São Marcos- Campinas- SP	**2000**	**2010**

Per capita income (in R$)	443,86	627,28
% extremely poor	2,79	2,33
% poor	11,37	7,04

Source: Adapted from UNDP, Ipea and FJP. 2016.

In 2010, the activity rate of the economically active population was 71.02% in the São Marcos neighbourhood. At the same time, the unemployment rate in the UDH (i.e. the percentage of the economically active population aged 18 and over who are unemployed) is 6.60 per cent (UNDP, IPEA; 2016).

Composition of the population aged 18 and over - 2010

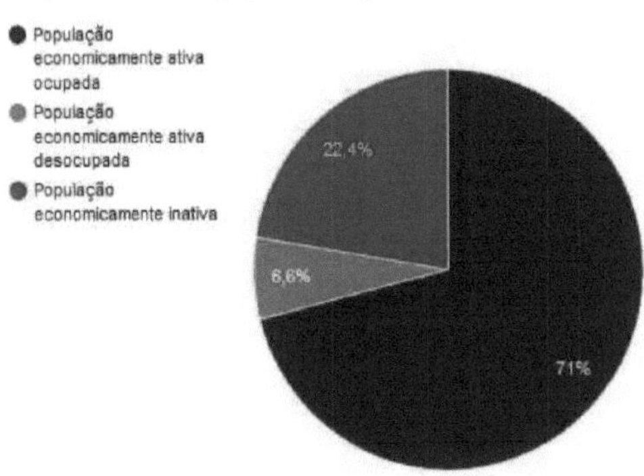

Source: UNDP, IPEA, FJP, 2016

Figura 7: Composition of the population aged 18 and over - São Marcos

Occupation of the population aged 18 and over	**2000**	**2010**
Activity rate	71,14	71,02
Unemployment rate	15,65	6,6
Degree of formalisation of the employed - 18 years and over	66,63	78,81
Educational level of the employed		
% of employed with complete primary education	44,73	67,13
% of employed with completed secondary education	20,76	43,76
Average yield		
% of employed people with an income of up to 1s.m.	18,72	8,47
% of employed people with an income of up to 2s. m.	70,9	66,94
Percentage of employed people earning up to 5 minimum wages	96,36	96,79

Source: Adapted from UNDP, Ipea and FJP. 2016.

After analysing all the MHDIs shown in this chapter, it can be seen that the initiatives are located on the outskirts of the municipality and had a low and medium development index in 2000, with the exception of the initiative located in Parque dos Eucaliptos, which has always had a high development index.

Poverty also follows a pattern: in most of the units, the average rate was 2.6% extremely poor, and in 2010 it fell sharply, to 1.4% in most neighbourhoods.

The differences between the initiatives are that some have improved more than others, such as Parque Eucalipto, which already had a high MHDI and is now very high, and Vila Brandina, which had a low MHDI and is now high in 2010.

In education, we see significant changes, for example, in Vila Brandina, the MHDI in 2000 was 0.391 (very low) and in 2000 it reached 0.685 (medium), the other neighbourhoods also increased, but proportionally, from low to medium, for example.

It's not the purpose of this article to assess the reasons for these differences, our aim so far has only been to characterise them and understand why there are community gardens in these neighbourhoods.

CHAPTER 4

A case study of HUP in Campinas: the Parque do Itajaí vegetable garden 3

The aim of this chapter is to present a case of urban gardening in Campinas, which corresponds to the Horta Comunitária do Parque do Itajaí 3 initiative.

The study was supported by the preceding discussion, presented in the previous chapters and contextualised within the other initiatives briefly presented in chapter 3.

In addition to secondary sources, fieldwork was carried out, with interviews on aspects of organisation, technology, community structure and coexistence, and the use of water on the plantation.

This fieldwork was carried out on 22 August 2016, and the content presented below was collected and systematised using the interview method, provided by Orlando Batista Santos, director of the Cio da Terra Association. The script from which the questions used to draw up the questionnaire were chosen is presented in the Annex.

4.1. Characterisation of the area of Initiative 5 - Itajaí Park 3

The Parque Itajaí 3 neighbourhood is part of the Campo Grande region and is located in the north-west of Campinas. According to the municipality, it is a region of strong economic expansion.

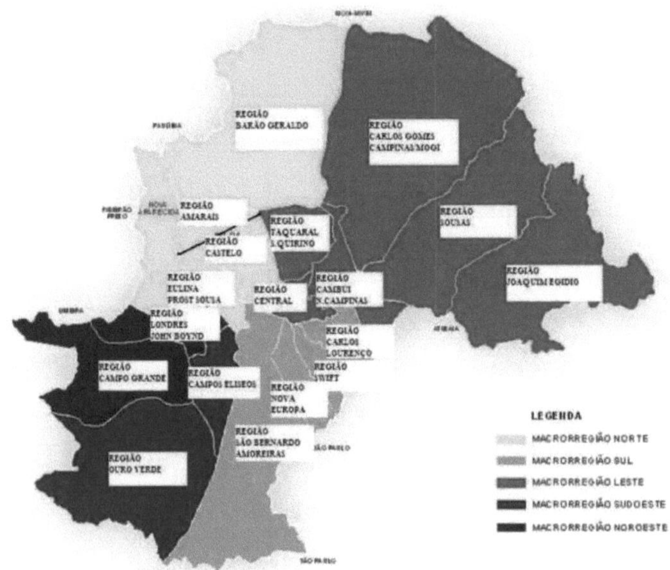

Figura 9: Macro-regions of Campinas
Source: Campinas City Hall (2016)

An aerial photo shows the site of the vegetable garden, located in Rua Doutor Pedro Miguel:

Figura 10: Location of the vegetable garden in the Parque Itajaí neighbourhood Source: Google Maps

The Municipal Human Development Index (MHDI) for this region in 2010 was 0.636, situated in the MHDI range of 0.600 to 0.699. The dimension that contributes most to the locality's MHDI value is Longevity, with an index of 0.747, followed by Income, with an index of 0.637, and Education, with an index of 0.540 (UNDP, IPEA, 2016).

In education, the proportion of children aged 5 to 6 in school was 95.08% in 2010. In the same year, the proportion of children aged 11 to 13 attending the final years of primary school was 86.02%; the proportion of young people aged 15 to 17 with completed primary school was 54.15%; and the proportion of young people aged 18 to 20 with completed secondary school was 22.92%.Analysing the region's income, we find a value of R$422.38 in 2010, while in the municipality of Campinas it was R$1,390.83 and in the Metropolitan Region of Campinas, R$1,148.94. (PNUD, IPEA, 2016)

Municipal Human Development Index and its components - Parque Itajaí 3

MHDI and components	**2000**	**2010**
HDI Education	0,322	0,54
% aged 18 or over with complete primary education	20,03	37,91
% of 5 to 6 year olds attending school	54,13	95,08
% of 11 to 13 year olds attending the final years of primary school	63,5	86,02
% of 15-17 year olds with complete primary education	34,16	54,15
% aged 18 to 20 with completed secondary education	11,27	22,92
MHDI Income	0,576	0,637
Per capita income (in R$)	288,08	422,38

Source: Adapted from UNDP, IPEA, FJP, 2016.

With regard to labour, in 2010, the activity rate of the population aged 18 and over (i.e. the percentage of this population that is economically active) was 68.87% in the region. At the same time, the unemployment rate in the same place (i.e. the percentage of the economically active population aged 18 and over who are unemployed) is 13.05%.Adding up the economically active population and the inactive and unemployed population gives a percentage of 31.2%, which justifies the presence of community gardens for both therapeutic and income-generating purposes, as shown in the graph below.

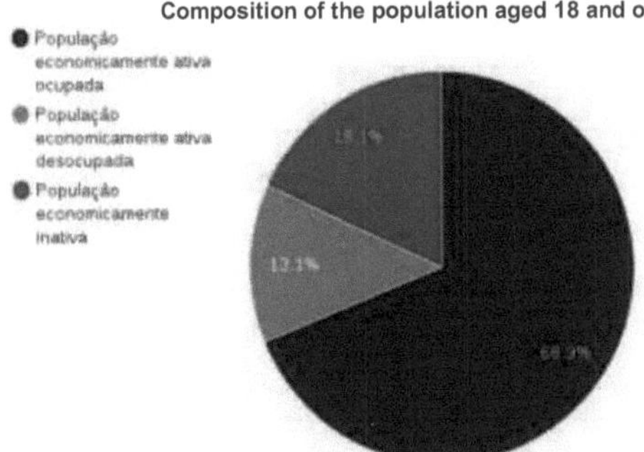

Figura 11: Composition of the population aged 18 and over - Parque Itajai
Source: UNDP, IPEA, FJP, 2016.

Another piece of data that confirms the use of community gardens in Parque Itajaí 3 is social vulnerability, given that 46.62% of mothers are heads of household and that, according to research carried out by Castelo and Alcântara (2011), women still assume a greater share of responsibility for caring for the home. The use of gardens is justified by the fact that they are close to homes, which facilitates the division of agricultural labour and household work.

Social Vulnerability - Parque Itajaí 3

Children and young people	**2000**	**2010**
Infant mortality	29,5	24,2
% of children aged 0-5 out of school	80,27	48,7
% of children aged 6 to 14 out of school	8,36	4,21
% of people aged 15 to 24 who do not study, do not work and are vulnerable, in the population of this age group	21,41	14,57
% of women aged 10 to 17 who have had children	7,22	3,01
Activity rate - 10 to 14 years	16,38	2,03
Family		
% of mothers who are heads of household without a primary education and with a minor child, out of the total number of mothers who are heads of household	26,29	42,62
% of vulnerable and dependent elderly	1,53	1,95
% of children up to 14 years of age with per capita household income equal to or less than R$ 70.00 per month	5,93	3,69
Labour and Income		
% vulnerable to poverty	54,99	34,87
% of people aged 18 and over without complete primary education	58,9	40,36

and in informal employment		
Housing Conditions		
% of population in households with bathrooms and running water	93,22	93,24

Source: UNDP, IPEA, FJP, 2016.

4.2. Initial breakdown of information on the HUP initiative in Campinas

The Parque Itajaí Community Garden initiative began in 2004, at the fork in the road that led to the implementation of vegetable gardens by Campinas City Hall, where many people were not served by the programme, which due to numerous bureaucracies ended up being an exclusionary and restrictive model because the land was not shared equally, and there was a lot of idle land below, so the community garden was born. It is on public land, granted by the City Hall for a period of 20 years, and came into effect in 2010.

Figure 12: Panorama of Horta Parque Itajaí
Credit: Leonardo Yvens

The oldest vegetable garden in the city of Campinas, however, is in Jardim Londres, located in Parque Eucalipto. It began in 1983 and was founded by Orlando Batista dos Santos, who is a member of the board of the Association of Producers of Urban and Rural Agriculture.

Periurban Campinas and Region[3] , and has been present in the Itajaí garden since its creation. The aim of the Itajaí initiative is to underpin agroecology.

> "The vegetable garden has a simplified process, we don't have such an expressive production, but every experience is valid, we want to support and stimulate **food production in the cities"** (Orlando, 60, 2016).

There are no community leaders, but there are people who have followed the development of the garden from the outset, such as the founder, Mr João Novais, who started growing vegetables on the site for occupational therapy; then other residents followed suit, forming a group of producers who were in compliance with current municipal legislation and became recognised by the public authorities. Mr João notes that his life changed with the vegetable garden:

> **"When the City Council launched the project, I realised it wasn't for me, so I had** this empty

[3] Associação dos Produtores da Agricultura Urbana e Periurbana de Campinas e Região - Cio da Terra - is a private non-profit organisation.

plot of land, got some people together and started planting. My family and I are depressed and the vegetable garden is a way of forgetting our problems; when my son died, **I brought my wife to help me plant"** (João Novais, 67, 2016).

In total there are around 32 members - this is an approximate number, since there is a significant turnover - who are spread throughout the neighbourhood, since there are a total of three gardens on the same plot of land, two community gardens and one from the City Hall project. In total, 11 families are assisted and benefited by Horta de Itajaí. There are many initiatives in this region, whether in public places or even in people's backyards; all of them are supported by Horta Itajaí 3.

Figures 13 (a) (b): View of Horta Parque Itajaí Credit: Leonardo Yvens

The total area set aside for the vegetable garden is 10,000 metres². The land is fertilised with chicken manure, composted with vegetable waste and liquid fertiliser and bio-fertilisers. The biofertiliser is made from crop residues that have to be chosen beforehand. There is a box on site where the material is placed, which then deteriorates and generates slurry. Mr Orlando monitors the pH of the compost, because he uses lime to make the biofertiliser in order to provide calcium for the soil, reducing its acidity.

Figure 14(a): Liquid fertiliser made with chicken manure Photo 14(b): Biofertiliser production site; on the right, Mr Orlando

Credit: Leonardo Yvens

Figure 15 : Measuring pH and dissolved nitrogen in the biofertiser. Credit: Leonardo Yvens

The farming carried out is organic, with a wide variety of vegetables, both leafy and fruit-bearing, as well as medicinal plants. Some of the products planted are: lettuce, almeirão, spinach, carrots, radishes, cauliflower and leeks. Some fruit is also planted, such as bananas and papaya, in order to extract potassium and fertilise the soil for its preparation. Below is part of a report on the Cio da Terra blog, written by Orlando dos Santos, which tells how the medicinal plant area has been received by the neighbourhood's population:

> **"People come to the** *green pharmacy* **every day** in search of a plant for therapeutic purposes, and the exchange of experiences between the nutritionist and the community is an important factor.
>
> "In general, people already know the type of plant they want. At the same time as learning about yet another novelty in the use of plants, the nutritionist [himself] also has the chance to advise on options that guarantee safety and ease in the use of the plant, or even recommendations to seek immediate medical help, depending on the justification **given."** (Orlando, 2013)

One of the problems the producers face is soil degradation; for this they have technical help from UNICAMP, especially at the start of the project, and also from EMBRAPA (Brazilian Agricultural Research Corporation) and CATI (Coordination of Integral Technical Assistance), with fortnightly meetings with the latter body.

When I spoke to the farmers, most of them who work there have lived in the fields and have a great deal of prior knowledge about plantations. Mr Tião's case bears this out. He currently lives in a flat, but when he was younger he lived in the countryside; for him, the vegetable garden is for his own consumption and for therapy. One example of a practice they already knew and implemented was composting; as the soil was practically sterile, composting helped to restore its fertility and other characteristics that made it suitable for horticulture.

Fotos 16 (a) (b) : Composteiras presentes na horta
Crédito: Leonardo Yvens

Photos 16 (a) (b) : Compost bins in the garden Credit: Leonardo Yvens

4.3. Water in the garden

The garden has a spring that is pumped to the plantation, but SANASA doesn't monitor the water for irrigation, which makes it difficult to certify the vegetables as organic.

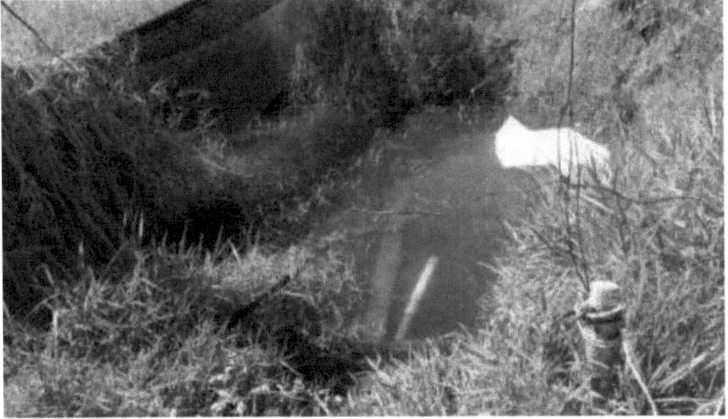

Figure 17: Spring from which water is drawn for planting Credit: Leonardo Yvens

4.4. Organisational knowledge in the initiative

There is no registration form for those who want to take part in the garden; in principle, anyone can take part and their economic situation is assessed[4] , after which they are given a small piece of land to plant

[4] The interviewee was not given the criteria or set of criteria for a new producer to join the initiative.

on.

When it comes to managing the garden, tasks are divided according to each person's physical ability, with the strongest being responsible for weeding, mending and also carrying the heaviest things, for example. The person who organises and controls this way of working, using a spreadsheet, is Mr Orlando, the director of the association, as already mentioned.

It's interesting to note that the Cio da Terra Association, represented by Mr Orlando, has set up a blog where they publish news about the vegetable garden in Parque Itajaí 3, which makes the project viable and accessible to the population.

The results of production are divided according to what each person has planted, and there may be an exchange of crops between them.

The surplus is commercialised in four ways: i) customers from the community themselves who pick up from the garden and have the opportunity to choose which vegetable they want; ii) sale to a vegetarian restaurant located in Barão Geraldo - called Raízes Zen; iii) through the "green exchange" fair that also takes place in Barão Geraldo every Tuesday; iv) and by delivery, for which the interested population registers and places orders, from which the producers deliver to the homes; and v) the surplus is donated to the Casa dos Anjos, which is located in the neighbourhood's Catholic Church and whose purpose is to assist underprivileged children.

The initiative has educational projects, and schools in the municipality are invited to visit the garden. Every Thursday a kindergarten class goes to the garden, and these visits are usually documented, as the photos collected on the blog attest. According to data from the Cio da Terra *blog*, in August the CEI Prof[l] Else Feijó Gomes school in the São Luis neighbourhood paid a visit, assisted by staff and a teacher, with the aim of establishing a partnership to exchange knowledge and support the project to set up a vegetable garden at the school.

Figures 18: (A) And (B) CEMEI Deputado Federal João Herrmann Neto, in Parque Itajaí IV; Source: Blog Cio da

As this is not a commercial initiative, there is no such thing as a "monthly income", although the vegetable garden project does make it possible for those who take part to supplement their income.

The biggest difficulty for producers, according to the interviewee, is keeping the area productive, since degraded soil makes it expensive to produce.

The profile of the people who work on the project are residents of Parque Itajaí 3. Most of them have no other form of income, being retired, unemployed and those seeking occupational therapy, and there is a very high turnover of participants, since when a person who was unemployed finds a job they don't stay in the garden.

The producers' children don't work in the garden, but they occasionally accompany their parents and help them plant and harvest.

The initiative has a longer-term vision, aimed at training teenagers from the community in planting and harvesting techniques, with a course in vegetable production, but nothing has materialised yet.

> **"The ideal thing would be to train these young people so that they could then go to** university to study agronomy/agroecology and bring their knowledge back to **us."** (Orlando, 2016)

At the end of his visit to the garden, Mr Orlando dos Santos summed up what the Community Gardens project means to him:

> **"Community gardens involve the community,** either consuming or producing food that will give people food security; that's the most important point. Then there's the educational aspect: the school will be able to interact with the production of vegetables and the children will learn from an early age where their food comes from, which is interesting. Other aspects are that retired people and people with health problems can be helped by working in these places, even the unemployed, but it's important to know that **urban vegetable garden spaces are** possible **and viable, and that they can generate income and employment."** (Orlando, 2016)

CHAPTER 5

Final considerations

In the light of what was proposed at the beginning of the paper, the aim was to identify and analyse community garden initiatives in the municipality of Campinas based on the concepts and principles of Solidarity Economy. Fieldwork was carried out in Parque Itajaí 3 in order to get to know and understand self-management practices - technical and organisational - and to analyse the principles of solidarity and cooperation, which end up re-signifying work practices.

The methodology used for this research was applied in two phases, the first based on an extensive literature review and mapping of vegetable gardens using secondary sources, and the second on fieldwork, elucidating the socio-economic aspects of the initiatives with HDI issues.

The concepts discussed in the paper seek to characterise how these initiatives are treated in the literature and what geography's perception is of the delimitation of space.

Due to a shortage of time, it was only possible to visit one vegetable garden, located in the Itajai 3 neighbourhood, but it is worth mentioning that this initiative serves as a model for others, as the founder of the first community garden in Campinas, Mr Orlando, director of the Cio da Terra Association, is responsible for managing the Itajaí 3 garden.

With regard to the concepts of urban and peri-urban agriculture, it is still a diffuse field, in which the authors seek elements to characterise the functioning and types of activities present, but do not portray, for example, their location in the cities/ countryside dichotomy, and it is up to an interpretation according to the area to be worked on, in this case based on a geographical view of the subject.

The literature does not reflect the great diversification of the Solidarity Economy in the case of urban horticulture, but a common point found in the reading is the possibility of community gardens being a model of self-management, based on issues of co-operation rather than competition.

With regard to these possibilities and difficulties, although the literature researched sheds light on the analysis of the social and production practices involved in urban and peri-urban community garden initiatives, there is a deontological difficulty, i.e. related to the lack of systematised research protocols to support field research. In addition, this study noted the need for in-depth information, especially on the social practices in these initiatives, which would require refining the research protocols and extending the research time, including the need for more technical visits and the inclusion of other actors as sources of information. An interdisciplinary approach, covering sociological, political and economic aspects, as well as analysing practices and technologies would be very useful.

Thus, one of the conclusions of this work points to the need for greater alignment between the theoretical approach and the methodological design. For example, the significance of carrying out a horticultural study refers to the real difficulties involved in implementing and organising the fieldwork: the dispersed nature of the initiatives in the municipality with consequent difficulties in transport, communication and information retrieval.

Within the scope of the study, we do not have the necessary information on the reasons why the results found for the vast majority of the improvements in the performance of the MHDIs for the human development units, in which the horticultural initiatives under study are located, can be evaluated. Particularly interesting are the improvements in the Education MHDI. Particularly noteworthy is the increase in the percentage of the population aged 18 and over, the % of children aged 5 to 6 attending school, the % of children aged 11 to 13 attending the final years of primary school, the % of young people aged 15 to 17 with completed primary school and the % of the population aged 18 to 20 with completed secondary school. It is improving, but less than the other indicators.

The analysis of the indicators led the research to focus on factors seen in the fieldwork, for example, the improvement in education is reflected in the actions of Horta Parque Itajaí 3, in which the presence of schools is constant in the environment, as well as the idealisation of a vocational course in horticulture for young people in the community.

For future work to be carried out, there are a few questions that could serve as support for the delimitation of research: Could community garden initiatives break away from a certain "welfare" profile and promote more integrative, active practices from the point of view of social, economic and political inclusion"? "Are there practices in these communities and associated with these initiatives that seek education in the broadest sense for those involved?" "What would be the elements for an "inclusive, participatory and citizen education" in the context of these initiatives?"

Another insight gained from the study was that the communities involved in these initiatives seem to have a changing composition over time, i.e. the presence of unemployed and elderly people who use work in the gardens as occupational therapy or a temporary way of generating income until they find another job. However, in an interview, the director of the association commented on the idea of a horticulture course for young people that the initiative intends to offer, with the aim of training permanent participants for the project.

Finally, this work is an additional element in understanding the structure of community garden initiatives in the municipality of Campinas and can serve as a basis for future work, as mentioned above.

The work reflected on a number of problems, but it was clear that field research was a necessary and entertaining tool, which was able to address and resolve doubts that were often missed in the literature. The principles of the Solidarity Economy were perceived in the initiative visited, in which not only in the form of work but also in social relations, mutual help and respect for the diversity of each individual, everyone's work was valued, which allowed for a full harmony.

Bibliography

AQUINO, Adriana Maria de; ASSIS, Renato Linhares. Organic agriculture in urban and peri-urban areas based on agroecology. Ambiente e Sociedade, Campinas, vol. 10, n. 01, pág. 137-150, jan/jun, 2007.

ARRAES, Nilson A M; CARVALHO, Yara M C. Urban agriculture and family farming: conceptual and practical interfaces. Informações Econômicas, SP, v. 45, n. 6, nov./dec. 2015.

ARRUDA, Juliana. Urban and peri-urban agriculture in Campinas/SP: analysis of the community gardens programme as a subsidy for public policies. Campinas, SP, 2006. p. 165.

BOULIANNE, Manon. Urban Agriculture and Development: The Mexican Experience. Nouvelles Pratiques Sociales, Volume 13, number 1, 2000. Available at < http://agriculturaurbana.org.br/sitio/textos/aumexico.htm>.

CASTELO BRANCO M; ALCÂNTARA FA. 2011. Urban and peri-urban gardens: what does the Brazilian literature tell us? Horticultura Brasileira 29: pp. 421-428.

JOB SCHMITT, Claúdia; TYGEL, Daniel. Agroecology and Solidarity Economy: trajectories, confluences and challenges. In: PETERSEN, Paulo (Org.) *Agricultura Familiar camponesa na construção do futuro.* Rio de Janeiro, 2009. P.105-127.

LEFEBVRE, H. The Urban Revolution. Belo Horizonte: UFMG, 1999.

LEFEBVRE, H. *De lo rural a lo urbano.* Barcelona: Penninsula. 1978.

MACHADO, Altair Toledo; MACHADO, Cynthia Torres de Toledo. Urban Agriculture. Plantina, DF, Embrapa Cerrados, 2002, 25 p.

MARTIN, Manuel A.Z. Agricultura urbana, condicion para el desarrollo sostenible y la mejora del paisaje. *Anales de Geografía,* vol. 35, n° 2, p. 167-194, 2015.

MONTEIRO, A. V. V. M. Agricultura urbana e Periurbana. In: *Revista de Agricultura Urbana,* 2007. Available at < http://agriculturaurbana.org.br/>. Accessed on 17 April 2016.

MOTTA, Eugênia de S. M. B. A "outra economia": um olhar etnográfico sobre a economia solidária. Rio de Janeiro: UFRJ, 2004. 102 p. Dissertation (Master's in Social Anthropology), Federal University of Rio de Janeiro, Postgraduate Programme in Social Anthropology, National Museum, 2004.

MOUGEOT, Luc J.A. Urban agriculture: concept and definition. Revista de Agricultura Urbana, n°1 , 2005. Available at <http://agriculturaurbana.org.br/RAU/AU01/AU1conceito.html>. Accessed on 14 April 2016.

UNITED NATIONS AGRICULTURE AND FOOD ORGANISATION - FAO .Urban and Peri-urban Agriculture in Latin America and the Caribbean: A Reality . Available at<

http://agriculturaurbana.org.br/sitio/textos/FUM%20IPES FAO.pdf>. Accessed on 6 May 2016

UNITED NATIONS FOOD AND AGRICULTURE ORGANISATION - FAO. Urban and Peri-urban Agriculture in Latin America and the Caribbean: A Reality . Availableat< http://agriculturaurbana.org.br/sitio/textos/FUM%20IPES_FAO.pdf>. Accessed on 6 May 2016.

UNITED NATIONS ORGANISATION FOR AGRICULTURE AND FOOD. Urban and peri-urban agriculture in Latin America and the Caribbean: a reality. Available at < http://www.fao.org/fileadmin/templates/FCIT/PDF/ Brochure_FAO_3.pdf>. Accessed on 6 May 2016.

NORTH-WEST REGION OF THE CITY OF CAMPINAS. Available at: http://www.campinas. sp.gov.br/governo/ servicos- publicos/regioes/noroeste/index.php.Accessed on 28 September 2016.

ROESE, Alexandre Dinnys. Urban agriculture. Available at < http://www.cpap.embrapa.br/publicacoes/online/ADM036.pdf>, Embrapa Pantanal, 2003. Accessed on 14 April 2016.

ROSA, Pedro P V. Public policies on urban and peri-urban agriculture in Brazil. Revista Geográfica de América Central, Costa Rica, special issue, p.1-17, 2011.

SANTOS, Milton. Technique, space, time. Globalisation and the Technical-Scientific Information Environment. 3. ed. Editora Hucitec: São Paulo, 1997.

SILVA, Anelino Francisco. The city-country relationship: how to analyse it? Imagem Gráfica e Imagem: Natal, 1998.

SINGER, P. Solidarity economy INTERVIEW WITH PAUL SINGER (2008). São Paulo: *Estudos Avançados.* Interview granted to Paulo de Salles Oliveira.

SPOSITO, M. E. B. The city-countryside question: perspectives from the city. In: SPOSITO, M. E. B; WHITACKER, A. M. Cidade e Campo: relações e contradições entre urbano e rural. 1. ed. Editora Expressão Popular: São Paulo, 2006.

TRANI, Paulo, PASSOS, Francisco Antônio; MELO, Arlete Marchi Tavares de; TRIVELLI, Sebastião Wilson; BOVI, Odair Alves; PIMENTEL, Eloísa Cavassani. Vegetables and Medicinal Plants: a practical manual. Campinas: Instituto Agronómico, 2nd revised edition, 2010, 72 p.

VILELA Sérgio L, O; MORAES Maria Dione C. Agricultura Urbana e Periurbana: Limites e Possibilidades de Constituição de um sistema agroalimentar localised no município de Teresina- PI.Teresina: *Revista Económica do Nordeste*, v. 46, p. s/n-s/a, 2015.

Sites Consulted

Community gardens and their distribution around the world. Available at http://fallingfruit.org/?locale=pt-br. Accessed on 30 October 2016.

IDHM Campos do Amarais. Atlas of Human Development. Available at <http://www.atlasbrasil.org.br/2013/pt/perfil_udh/29498>. Accessed on 7 October 2016.

HDI of the Campo Grande Region. Available at <http://www.atlasbrasil.org.br/2013/pt/perfil udh/29488>. Accessed on 27 September 2016.

IDHM Jardim Londres. Atlas of Human Development. Available at < http://www.atlasbrasil.org.br/2013/pt/perfil udh/29524>. Accessed on 27 September 2016.

IDHM Santa Rosa. Atlas of Human Development. Available at <http://www.atlasbrasil.org.br/2013/pt/perfil_udh/29282>. Accessed on 7 October 2016.

IDHM São Marcos. Atlas of Human Development. Available at <http://www.atlasbrasil.org.br/2013/pt/perfil udh/29464>. Accessed on 7 October 2016.

IDHM Vila Brandina. Atlas of Human Development. Available at <http://www.atlasbrasil.org.br/2013/pt/perfil_udh/29545>. Accessed on 7 October 2016.

Annex

Urban and Peri-urban Horticulture in Campinas: an integral tool for collecting information

This set of questions was drawn up on the basis of aspects raised in the literature for application to those involved in community garden projects in the city of Campinas. It consists of 53 questions grouped into six parts, namely: i) general questions about urban and peri-urban horticulture (UHP) in Campinas; ii) detailed information about UHP initiatives in Campinas; iii) technical knowledge about the initiatives; iv) organisational knowledge; v) water in the garden; and vi) the community beyond the garden. This set of questions served to inspire the itinerary of visits and questions to be applied to those responsible for the gardens selected for the case studies.

Brunna D'Luise Turato Lotti Alves

Full version based on which questions were selected for the case studies

Part I. Urban and peri-urban horticulture (UHP) initiatives in Campinas: general

This section lists the questions that should allow for a general identification of the HUP initiatives in Campinas.

1. Mapping the network and describing the actors involved: their roles in setting up and implementing urban and peri-urban agriculture initiatives
2. What are the initiatives? (MAPPING - SPACIALISING - use ArcGis for this)

Part II. Initial breakdown of information on HUP initiatives in Campinas

This section lists the questions that should enable the specific identification of each of the HUP initiatives in Campinas. These are questions to be asked during visits to the gardens. The name of the contact (the person who will be interviewed and who will answer these questions) should be obtained in advance by e-mail or telephone. A copy of the questions below should be sent to the respondent by e-mail. The questionnaire should also be printed out and three copies taken on the day of the interview.

3. Identification of garden sites (including data for use in ArcGis)
4. Start date of initiatives
5. Identifying a "leader" or a group of "leaders" in the initiative
6. How many (and which) families are involved in the project?
7. What activities did they do before?
8. What are the stated objectives of the project?
9. History of initiatives
    - "How did it all start...?"
    - Key facts, figures, institutions and dates in the "history" of the initiative
10. What area (m^2) does the initiative occupy?
11. What are the types of crops? (What are the varieties of plants?)
12. How much is produced (by type of crop, using units of account such as Kg, "feet of..." etc)
13. What is your production used for? (Own consumption, sale at the local market, sale at CEASA...)
14. What is the "meaning" of the project, in the view of its gardeners?

Part III. Expertise in HUP initiatives in Campinas

In this section, the questions that will be used to identify the nature and origin of the knowledge used in each of the initiatives are listed. We are interested both in identifying information about formal (technical-scientific) knowledge and the traditional knowledge that makes the gardens possible.

15. At the start of the project, what was the previous knowledge of horticulture among the people involved?
16. Is there support from an agricultural technician or agronomist? Which institution (city hall, IAC, CATI, other)?
17. Do gardeners attend any training courses on planting or management techniques? If so, who offers the course? And how are the gardeners selected to take part in the course?
18. What are the techniques for preparing the land? What about looking after the crops?
19. Do you fertilise the beds?
20. When and what is it used for?

21. Do you carry out a soil analysis before fertilising? Which institution does this analysis?
22. Are there any pests (insects and/or others)? Which pests?
23. How are insects and other pests controlled?
24. How are other diseases controlled?
25. How do you control weeds?
26. What agricultural tools / implements are used in this garden? How are they acquired? How is it maintained?
27. What agrochemical inputs (fertilisers, herbicides, fungicides and others) are used in this garden? How are they acquired? Where are they purchased? How often are they purchased? How are they stored? How are they stored? What is the destination of the packaging?
28. What "natural" inputs are used in the garden (Bordeaux mixture, Neen extract, other natural products)?
29. What are the other inputs?

Part IV. Organisational knowledge in HUP initiatives in Campinas

In this part, there are questions about the management, organisation and administration of work, resources and production in urban and peri-urban horticulture initiatives in Campinas.

30. Is the garden registered with the town hall or a specific public body (inspection)?
31. How is the garden managed? Is there a division of labour, such as buying inputs, controlling the production schedule (planting, caring for the crops, harvesting, etc.), recording and controlling costs, selling or sharing, etc.? How is this division done? Who is responsible for which of these tasks?
32. How are the results of production divided up? How are they distributed?
33. Are they commercialised? How? Where? Where?
34. What is your monthly income from produce sold in the garden?
35. Where do you buy vegetable seedlings or do you produce your own? How are they acquired?
36. Where are vegetable seeds sourced? How are they acquired?
37. Is there any action to certify production? What exists in this regard?
38. Are taxes paid? Which taxes?

Part V. Water in the garden

This part deals with questions about the water used in vegetable gardens: its sources, its uses, ways of controlling its quality and the destination of waste water.

39. What sources of water are used in this project? Do you use mains water or other water (if other, which)?
40. Water quality analyses: are they carried out? By which institution? How often?

41. How much water is used (monthly / annually) (if this information is available)
42. What is the (monthly / annual) cost of water for this project?
43. How is payment made for the use of this water? (Is there a provision of money for this? How is this provision made?)
44. Do you use irrigation? What kind?
45. Do you use water to wash vegetables after harvesting? Where do you wash? (tank, water tank, straight from the tap)
46. What other water uses are there in the project?
47. What is the destination of the waste water? Is there any form of reuse and which one(s)? Is there any form of treatment and which one(s)?

Part VI. The community beyond the garden

In this part, the aim is to understand aspects of the structure and conviviality of the community involved in the project.

48. Who are they?
49. Where do you live?
50. What other activities do you do? (e.g. do you do other work? What and where? Do you study? What and where?)
51. How are the children looked after in terms of education?
52. How and where do they receive health care?
53. What are the community's other forms of socialising? (Examples: churches, festivals, games, support in caring for children and the elderly...)

Printed by Books on Demand GmbH, Norderstedt / Germany